Life in the Desert

Written by Andrew Clements

STECK-VAUGHN
COMPANY

A Division of Harcourt Brace & Company

www.steck-vaughn.com

Contents

Desert Facts	3
What Makes a Desert?	6
The Land in the Desert	8
When the Land Is Dry	12
Desert Animals	14
Desert Plants	23
Deserts Around the World	26
Visiting the Desert	31
Glossary	32
Index	33

Desert Facts

What is a desert? It is a dry, sandy, dusty land area. It has very little water and gets very little rain. The desert can get really hot during the day. Not many plants can grow in a desert.

Deserts have cactus plants, snakes, and a few kinds of flowers. Some deserts also have mountains and **valleys**. Deserts are full of beauty and have many surprises.

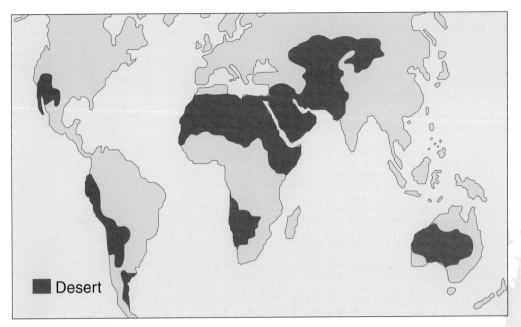

Deserts are found all around the world.

Where are the deserts? They are all over the world. Deserts cover about one third of Earth's land. The largest desert in the world is the Sahara. It covers almost one third of Africa.

There are no deserts in the lands that are far north. These areas have a great amount of snow, ice, and rain. That keeps the land from drying out.

How hot are deserts? Most deserts have a **temperature** of about 94°F (34°C) in the daytime. That is pretty hot, but people can still go outside for a short time. But at night, the desert cools down to about 32°F (0°C). That is freezing.

How dry are deserts? Deserts are very dry. Most deserts get less than ten inches of rain each year. In Chile, there is a desert where no rain has fallen for 400 years. It is so dry that few plants grow there.

This desert in Chile may be the driest in the world.

What Makes a Desert?

When the land gets too little rain, a desert forms. An example is in the deserts of North America. Wet winds blow in from the ocean. Wind that blows over the ocean makes clouds. The wind blows the clouds toward the mountains. When the clouds hit the mountains, rain falls.

This is why it rains on the side of the mountains that is near the ocean. But there is no rain on the other side of the mountains. The winds that blow there are dry winds. They have few clouds. Very little rain falls on the desert side of the mountains.

Dry winds blow in the desert.

The Land in the Desert

Sometimes it does rain. A sudden rainstorm can make a flood in the desert. Rushing water picks up sand and rocks. This rushing water can cut away the land.

An **arroyo** is a deep cut or gully in the desert land made by rushing water. There is sand at the bottom. The sides are steep.

This desert arroyo was made by rushing water.

Over time, an arroyo can become a **canyon**. Canyons are larger and deeper than arroyos. Canyons are cut by streams and rivers.

Rivers can cut very deep canyons. The most famous canyon is in the United States. It is called the Grand Canyon. The Grand Canyon was cut by a mighty river.

Many visitors go to see the Grand Canyon.

There can be strong winds in the desert. The wind picks up the dry soil and sand. The wind blows the sand at rocks and hills. Over a long time, the blowing sand can cut rock. A tower of rock can be cut by the wind. It is called a **butte**.

A butte in Arizona was built by blowing sand.

Blowing sand can cut beautiful shapes. This happens over many, many years, little by little.

Sand **dunes** are made when sand piles up in mounds. In Saudi Arabia, there is a huge area of sand dunes.

Sand dunes change with the wind.

When the Land Is Dry

When the land is dry, few plants can live in it. There are few roots to hold the soil. Wind can pull the plants up by the roots. Dry soil can blow away. There are sand dunes in some places and bare rocks in other places. When the wind blows hard, sand and dust fly through the air. A sandstorm can be blown thousands of miles and damage crops.

Sand dunes change shape as the wind changes.

It is very hard for plants and animals to live in a desert. The hot sun beats down on the ground in the daytime. The rocks and sand get very hot. Then at night, the land cools down quickly. Most plants and animals can't get used to such big changes in temperature.

A desert iguana hides from the hot sun.

Desert Animals

Most desert animals hide from the hot sun. They dig burrows under the ground or hide under rocks. They stay hidden all day.

At night, the desert is cooler. Then most **mammals**, such as ground squirrels, foxes, and antelopes, come out to find their food.

A kangaroo rat lives in the desert.

Some mammals eat plants. Some mammals eat other animals. Most desert mammals are small. Big animals need a lot of food, but there is not much to eat in the desert. Mammals that eat other animals have to hunt for them.

A coyote hunts for food in the desert.

Rattlesnakes have poison.

There are snakes and lizards living in most deserts. Like the mammals, most snakes and lizards stay out of the hot sun. They come out at night to hunt. Some snakes, like the rattlesnake, are dangerous. They have poison in their bite to help catch food.

The Gila monster is a poisonous lizard.

One kind of desert lizard in North America has poison. It is called the **Gila monster**. It lives in the Sonoran Desert.

Most snakes and lizards don't drink water. They eat insects and small animals. They get water from the animals they eat. This helps them stay alive in the desert.

There are spiders of all sizes in the desert. The biggest spider is the **tarantula**. There are also scorpions in the desert.

Many insects, such as ants, termites, and bees, live in the desert. Most desert insects live underground. Some dig tunnels. Others hide under rocks. Most only come out of the ground at night.

The tarantula is the biggest desert spider.

Desert insects are very important. Insects such as bees and flies go from flower to flower. They help the plants grow.

Ants and termites eat trees and plants. Then they become food for other desert animals.

Insects like these termites are eaten by other animals.

The roadrunner is a desert bird.

Desert birds must look hard to find their food. Some eat insects, and some eat lizards. Others eat cactus fruit. Most desert birds only get water from the food they eat.

Some desert birds make nests on the ground or in low bushes. Others make their nests in cactus plants.

Some frogs can live in the desert.

Frogs and toads sleep under the desert sands. The first rains of the summer wake them up. They lay eggs in pools of still water. Eggs become tadpoles and then frogs or toads.

Desert tortoises eat cactus, grasses, flowers, and fruit. Their bodies store the water from food. Their hard shells keep them safe. A desert tortoise can live for many years.

The desert tortoise eats cactus.

Desert Plants

Not all deserts have cactus plants. Parts of some deserts are too cold for them to grow there. Most cactus plants will die if they freeze.

The Joshua tree is a kind of cactus that is a tree. It can live in the cold parts of the desert.

The Joshua tree needs little water.

Many kinds of cactus grow in the desert in Mexico. Over 200 kinds can be found. They store water, so they can live most of the year without rain. They often serve as homes to animals.

The **saguaro cactus** grows only in the North American desert. It is amazing because it can grow up to 60 feet tall.

The giant saguaro cactus can grow to 60 feet tall.

The **mesquite** tree grows in the desert. The mesquite is a small tree, but it has very deep roots. One kind grows curly pods with seeds in them. The pods are used as food by desert animals.

Desert animals and plants are able to make good use of desert lands. They also make the deserts beautiful.

Mesquite trees have deep roots to reach water.

The Sahara is the world's largest desert.

Deserts Around the World

The largest desert in the world is the Sahara. It is in Africa. This desert is about as large as all the United States put together. The Sahara gets about eight inches of rain a year.

In the mountains nearby, there is more rain and some snow. Some water flows from springs in the ground. Villages are found near the springs.

The Great Basin desert has mountains and valleys.

The Great Basin is a large desert in the United States. It is made up of many high mountains and low valleys.

The mountains in the Great Basin get rain sometimes. Trees can grow on the mountains. But the valleys are almost always dry. Only small plants can live there.

The Mojave Desert in California is really dry, even for a desert. It gets less than six inches of rainfall each year. Death Valley is part of the Mojave Desert. It is one of the hottest and driest places on Earth. In summer, it gets hotter than 120°F (49°C). That's in the shade!

Death Valley is in the Mojave Desert.

The Sonoran Desert in Arizona is the hottest desert in North America. Still, the Sonoran Desert has many trees and plants. The desert trees and plants live on little water.

Rains come each summer in the Sonoran Desert. The plants soak up the water and save it to use during the many dry months.

The Sonoran desert is full of plant life.

The Chihuahuan Desert is very large. Most of it is in Mexico. This desert also gets most of its rain in summer. Low bushes and cactus plants grow in the Chihuahuan Desert.

The Chihuahuan Desert in Mexico is rocky and dry.

Visiting the Desert

Would you like to visit the desert? If you went, what would you take? Take some sunscreen to protect your skin from the sun. Wear a hat to protect your head. You'll need a lot of water, too. Bring a camera to take pictures of the plants and animals.

The desert is a very special place. You'll see its beauty, its plants and animals, and the special shapes of the land.

These people are visiting Death Valley.

Glossary

arroyo a cut in the land made by water

butte land that juts out from wind cutting it over time

canyon a large cut in the land made by rivers

dunes piles of sand heaped up by the wind

Gila monster a poisonous lizard

mammals animals whose body temperature stays about the same

mesquite a small desert tree with deep roots

saguaro cactus the largest cactus plant

tarantula a very large, hairy desert spider

temperature how hot or cold it is

valleys low land between hills or mountains